Frances L. Strong

All the Year Round

Part III

Frances L. Strong

All the Year Round
Part III

ISBN/EAN: 9783337026134

Printed in Europe, USA, Canada, Australia, Japan

Cover: Foto ©berggeist007 / pixelio.de

More available books at **www.hansebooks.com**

ALL THE YEAR ROUND

A NATURE READER

PART III: SPRING

BY

FRANCES L. STRONG
ST. PAUL TEACHERS' TRAINING SCHOOL

ILLUSTRATED BY

GERTRUDE A. STOKER
TEACHER OF DRAWING, ST. PAUL

BOSTON, U.S.A., AND LONDON
GINN & COMPANY, PUBLISHERS
1896

NOTE TO THE TEACHER.

———◆◇◆———

I T is not the purpose of the author of this series to offer,
or even suggest, any rules for its use. If anything is
established in education, it is the fact that aside from
certain underlying principles and general directions, each
teacher must be a rule unto herself. The methods which
the author and her colleagues have found successful might
be entirely out of harmony with an equally good system in
some other city. It is to be presumed, however, that if this
series of nature-stories should be so fortunate as to be
received with favor by the educational public, it will occasionally find its way into the hands of some teachers who
are not familiar with nature-work as developed in large
cities, and well-organized school systems. To these it may
be interesting and helpful to know just "how it has been
done" in the schools out of which these stories grew, and
in which they have been used. Indeed, by way of com-
parison and suggestion, it may also be of assistance to those
who have passed through the experimental stage and have
wrought out a system of their own.

It has been the custom in the St. Paul public schools to
pursue the following plan :

Materials. — The teacher goes out with her pupils to
collect the materials referred to in the lessons, gathering

enough to allow each pupil one specimen. Animals and plants are kept alive in the schoolroom to enable all to study their growth and habits.

After the material is at hand, the development of a specific lesson is divided (though not formally and rigidly) into five parts.

I. *Morning Talk.* — The work of the day is begun with a morning talk based either upon one of the natural objects, or upon a geographical topic, according to the season.

If an animal, a plant, or a stone be the subject of the lesson, pains are taken to see that each child is provided with a specimen. By skillful questioning, statements are drawn from the children concerning the facts the teacher wishes observed. New words are occasionally suggested and written upon the blackboard, and their frequent use is required throughout the lesson. In studying objects, it has, of course, been found advisable to consider them as belonging to some great family, making comparisons, and finding resemblances and differences. Children readily find this family element in all things studied.

II. *Drawing.* — The observation lesson is followed by a drawing lesson upon the subject studied. The child has already been supplied with the plant or animal. Each child draws his specimen carefully. It is by no means necessary for the teacher herself to be able to draw in order to get results. Each child is simply required to reproduce with his pencil just what he sees, just as he sees it. Children illustrate their language papers on flowers with water-colors or pencil. Work in free-hand cutting can be given from all objects, such as bottles, leaves, animals, etc. Scissors are used for this cutting. Modeling in clay is done from any object that will correlate with the other work. It has been found that in connection with the myths there is a great opportunity to develop imagination by allowing the child to illustrate the stories.

III. *Spelling.* — A spelling lesson upon the new and difficult words will follow.

IV. *Reading.* — The child is now ready for the reading lesson appropriate to the subject.

V. *Language.* — Finally, the children write descriptions of the object or country studied, giving free expression to the facts each has acquired.

It may be added that great interest may be excited by introducing into the number-work problems concerning the subject of the morning talk.

The literature, also, holds a very prominent place in this nature-work. The following list suggests poems to be committed to memory, and stories to be read in connection with this reader :

PLANTS.

The Dandelion	*Hiawatha*
The Dandelion	*Parts of Lowell*
Shall I Go and Call Them Up? .	. *Susan Coolidge*
Legend of the Cowslip . . .	. *Sara Wiltse*
Elderberry Bush	*Hans Andersen*
The Fern	*Harper's Second Reader*
Fairy Land of Flowers	*Mara Pratt*
Little Flower Folks . .	*Mara Pratt*

TREES.

Published by A. FLANAGAN, Chicago.   Price, 20 Cents.

As will be inferred from the method outlined above, the purpose of this book will be entirely misconceived, if it is looked upon merely as a convenient means of furnishing new reading matter for children (although it is sincerely hoped that it will do this). It is intended also to stimulate the thought, enlarge the vocabulary, and open the eyes of the children to the wonders of the world around them.

In the St. Paul public schools the manuscript of this series has been used in the second grade. It is thought, however, that it may be used in the third, and even the fourth, with equally good results.

October 17, 1895.

TABLE OF CONTENTS.

———•◦•———

PLANTS.

viii*Table of Contents.*

BIRDS.

PAGE

25. THE ROBIN (Perching) 56
26. HOW THE ROBIN GOT HIS RED BREAST . . 59
27. WHAT ROBIN TOLD (Poetry) 62
28. SPRING IN THE APPLE TREE . . 63
29. THE HUMMING BIRD . . . 67
30. THE WOODPECKER (Climbing) . 70
31. THE ORIGIN OF THE WOODPECKER . 73
32. THE DUCK (Swimming) . 76
33. THE HEN (Scratching) 78
34. THE SNIPE (Wading) . . . 81
35. JACK AND THE OSTRICH (Running) . . 83
36. THE OWL (Prey) 87
37. THE PIGEON AND THE OWL (Poetry) . 90

ANIMALS.

38. THE FROGS' EGGS 92
39. FROGS AND TOADS . . . 94
40. THE TURTLES . . 96

SPRING.

SPRING.

1. PUSSY WILLOW.

ALL winter long Mother Nature had sent her winds to rock the babies' cradles. Back and forth they had rocked, lulling the baby buds to sleep.

Mother Nature did not want them to waken, for Jack Frost was king outside. She feared that her babies would freeze.

In the fall she had tucked them up snugly and given them a warm coverlet to keep out the cold. So the baby buds had slept soundly all winter.

At last it was time to awaken. Mother Nature sent her March winds to rouse the little buds.

There was another king of the air. The beautiful golden sun was king now.

Mother Nature knew that the new king loved the babies. He would help them to grow large and strong.

She told this king where the children were, and asked him to waken them. So one bright morning, the king sends down his warm rays. They tap on the brown covers; they send the sap up into the buds.

Soon Mother Nature hears calls from the cradles. She looks and finds that her babies are stirring. How glad she is!

Soon the little pussies ask, " May we come out ? "

" Yes," says Mother Nature, " if you will wear your fur hoods. It is very cold out here."

Some of the buds wish to go with nothing on their heads, but at last they all put on their gray pussy hoods and come out. At night it is so cold that they are very glad they wore them.

What a time they have rocking on the brown branches. Up and down, here and there, they rock in the bright sunshine. How many brothers and sisters there are!

King Sun is a good friend to them. He helps them every day to grow longer and longer. These pussies are called catkins.

2. PUSSY WILLOW.

"OH! you pussy willow,
   Pretty little thing,
Coming in the sunshine
Of the merry spring.
Tell me, tell me, pussy!
For I want to know,
Where it is you come from,
How it is you grow."

"Now, my little children,
   If you look at me
And my little sisters,
   I am sure you'll see
Tiny little houses,
Out of which we peep,
When we first are waking
From our winter's sleep.

"As the days grow milder
   Out we put our heads,
And we lightly move us
   In our little beds ;

And when warmer breezes
Of the springtime blow,
Then we little pussies
All to catkins grow."

Elinor Smith's *Songs and Games for Little Ones.*   OLIVER DITSON, Publ.

3. THE LILAC TWIG.

THE little lilac buds had been
waiting a long time for Mr.
Southwind. Mother Nature had told
the buds that when Mr. Southwind came
along, they might come out.

All winter, these little buds had been kept
warm. They had warm brown scales, which
kept out the cold.

These scales turn into leaves, but they were
hardened to protect the buds.

After Mr. Southwind came, the buds began to
swell. They grew larger and larger. They unfolded,
and the green leaves came out.

What do you think was in some of the buds
with the leaves? A flower bud.

One bud grew on one side of the twig, and
another grew just across or opposite. So we say
that the lilac buds grow opposite. Do all buds
grow in this way?

The lilac has two buds at the end of the twig.

The twig, on which these buds grew, had a light

brown coat outside.   Inside this brown coat was
a green one.   Then there was a third coat, which
was white.   The brown coat, with its linings of
green and white, made the bark of the twig.

Do you know what was inside these coats?
There was the hard wood, and in the center, the
soft pith.

There were many little white spots on the brown
coat.   Through these tiny openings, the twig gets
some air.   They are called breathing pores.

4. THE LIMA BEAN.

MY name is Lima Bean. I have come to show you what my kind old mother has done for me. She has given me a heavy white coat to keep me warm. If you soak me in water, you will find a pretty ruffle around my coat.

If you pull this coat off, I shall break 'into two pieces. These pieces are my two seed-leaves. There is sweet food packed in them. There is so much of it, that I make good food for man.

I need food, just as you do. You have to get and cook your food. I do not have to move to get mine.

My food is used to help the stem, the leaves, and the root of the plant to grow.

You will see a little white stem and two leaflets on one of my seed-leaves. These little leaflets are darker than my seed-leaves. Don't you think they are a queer color for leaflets? I wonder if you can tell why they are not green.

If a bean is put into the ground, the dampness of the earth gets into it. This makes the seed swell and the coat burst.

Then out comes a little root which pushes down
into the ground.   Do you know why the root
comes first?

Soon a little stalk comes out of the seed.   This
shoots upward, carrying the seed-leaves with it.

When they are out of the ground, the seed-leaves
turn green.   But when they have given the plant
the food it needs, they wither and fall off.   The
seed-leaves have helped the plant to grow much
faster and stronger than it could have done without them.

5. THE HORSE-CHESTNUT TWIG.

"GEORGE, do you know who I am?"

"Yes," answered George, "you are a horse-chestnut twig. Where did you come from?"

"My mother is a large horse-chestnut tree," said the twig. "She is very proud of her children in the springtime. They look so pretty in their green, white, and pink dresses."

"I have been looking at your buds," said George, "and I think you have been well cared for."

"Yes, all the buds have thick scales. When the weather gets warm enough, the bud pushes the scales apart. Have you noticed that a varnish is spread over all the scales?"

"Yes," answered George, "what is the use of this varnish?"

"The varnish fills up all the cracks. It helps to keep us warm, and keeps out the bugs and worms. If you should pick off the varnish and scales of one of the buds in the winter, it would die."

"I have never thought of that," said George. "Thank you, little twig, for telling me."

" If you look at me," said the twig, " you will find the leaf scars. These mark the places where the leaves were last year, and on these little scars you will find tiny dots. These will tell you the number of leaflets which made up the large leaf. Some have five, and some seven."

" What are those little rings around you? " asked George.

" They show how much I have grown each year," answered the twig.

" I hope I am not tiring you, little twig, but I have one more question that I should like to ask."

" If I can, I shall be glad to answer it," said the twig.

" Where are your leaf and flower buds? "

" Under my brown coat are my tiny little leaflets, all wrapped in warm cotton, waiting for spring. In between the leaflets, my flower bud is hidden. When the sun has helped to open my big leaf bud, the flower will come out in beautiful pink and white blossoms."

" Thank you," said George, " I am so glad I had this talk with you."

6. THE ELDERBERRY BUDS.

"HOW warm the sunshine is!" exclaimed a little bud on an elderberry twig.

"Yes," answered her twin sister, "I believe spring has come at last. How glad I am! Now we can have a peep at the world."

"How long the winter has been!" said the first little bud. "I am glad we live opposite. Let us open our doors, little sisters."

"No wonder we were warm all winter. See how well we have been cared for, with these thick walls to keep out the cold."

"Oh, sister," said the first bud, "here is a place just below me where a leaf grew last summer."

"I have one. They are leaf scars. Our twig has dots and rings on it, too."

"I have leaves and a flower bud," said the first bud. "The flower bud is made up of many little buds. It is green now, but a little later there will be many white blossoms."

"I have no flower buds in my house," said another.

"Some of the buds have leaves. Others have both flower buds and leaves," said one of the twins.

"Does any one know what comes after the flowers?" asked the first bud.

"I do," answered a robin near by. "A berry grows where you see each little flower. For some time these berries are green. The warm sunshine helps to ripen them."

When the berries are ripe, the bushes are very beautiful with their loads of red fruit. The berries of some elderberry bushes are black.

What a feast the birds have when these berries are ripe! But the birds do not get them all; people gather many of the berries to make them into sauce and pies.

7. THE BEAN PLANT.

YOU see the baby bean has wakened and been hard at work. We can see the roots, stem, and leaves very plainly now. Its seed-leaves have turned brown and fallen, for they are no longer needed.

This little plant must eat, drink, and breathe, just as you or I must. The roots, stem, and leaves do this work.

The poor roots work very hard. They reach their slender fingers about for food. They have to bring water from the ground for the whole plant.

This is no easy task, for the plant is very thirsty and seems to have no mercy upon the poor roots.

Another great duty of the roots is to hold the plant firmly in the ground. If anything should destroy the roots, the little plant would die.

The stem carries the food and water to the leaves and flowers. It is very careful of this food or sap.

The leaves have two great duties, to eat and
to breathe.   How do you suppose the leaf does
this eating and breathing?

Of course, you say it must have a mouth.   It has
not one mouth only, but hundreds of mouths.   You
cannot see these mouths without a microscope.
Most of them are on the underside of the leaf.

Have you ever thought how much comes from a
single bean?   You put a bean into the ground.   A
vine grows from it.   It blossoms.   Then come the
pods, and in the pods are beans like the one you
planted.

8. THE STRAW, THE COAL, AND THE BEAN.

ONCE upon a time, there was an old woman who lived in a village. One day she went into her garden to gather some beans for her dinner.

She had a good fire, but to make it burn more quickly, she threw on a handful of straw.

As she threw the beans into the pot to boil, one of them fell on the floor not far from a wisp of straw which was lying near.

Suddenly a glowing coal bounded out of the fire and fell close to them. They both started away, and exclaimed, " Dear friend, don't come near me till you are cooler. What brings you out here?"

" Oh," replied the coal, "the heat made me so strong that I was able to bound from the fire. Had I not done so, my death would have been certain, and I should have been burned to ashes by this time."

"Then," said the bean, "I have also escaped being scalded to death, for had the old woman put me in the pot with my comrades, I should have been boiled to broth."

" I might have been burned," said the straw, " for all my brothers were pushed into the fire and smoke by the old woman. She packed sixty of us in a bundle, but I slipped through her fingers."

" Well, now, what shall we do with ourselves? " asked the coal.

" I think," answered the bean, " we may as well travel away together to some more friendly country."

The two others agreed to this, so they started on their journey.

After traveling a little distance, they came to a stream over which there was no bridge. They were puzzled to know how to get over to the other side.

Then the straw said, " I will lay myself across the stream, so that you two can step over me, as if I were a bridge."

So the straw stretched himself from one shore to the other. The coal tripped out quite boldly on the newly-built bridge. But when he reached the middle of the stream and heard the water rushing under him, he was frightened. He stood still and dared not move a step farther.

Then a sad thing happened. The straw was scorched in the middle by the heat still in the coal. It broke in two from the weight of the coal and fell into the brook.   The coal, with a hiss, slid after it into the water.

The bean had stayed behind on the shore.   When she saw what had happened, she laughed so hard that she burst.

She would have been worse off than her comrades had not a tailor come to rest by the brook. He noticed the bean, and, being a kind-hearted man, he took a needle and thread out of his pocket.

Taking up the bean, he sewed her together.   She thanked him very much.

He had only black thread with which to sew the bean, so ever since that time some beans have a black mark down their backs.

From *Grimm's Fairy Tales.*

9. THE PEA VINE.

SOME time ago I was a little round pea. I had a coat and a baby plant that was fast asleep inside the seed-leaves.

I was put into the ground, and strange things happened to me.

I am held in the ground by my roots. They are not so large now as they will be after awhile, for now they have to hold and feed a small plant.

While they are traveling under the ground, my roots gather food for me. Perhaps you will wonder what has become of the food that was stored in my seed-leaves. Can you not guess?

Did I hear some one say that I ate it?

Yes, that is just what I did. I had just enough food given to me to keep me alive till my roots could reach out and gather it from the earth, and the leaves from the air. They have done good work, too, for you see how strong and healthy I look.

My leaves are heavy, and the stem is not strong enough to hold them up. I have some little tendrils which help to hold up the vine.

A voice said to the little tendrils, " Just as soon as you come to anything around which you can twist, do so. In this way, even if you are little, you can help your plant."

The little tendrils were very obedient, and so anxious to help that they began to twist around some dried twigs that the gardener had put there for me.

When the pods have grown, they will hold many peas sitting in a row. Then the tendrils will have to hold up a heavy load. But they will twist around the twigs so many times that the pea pods will be quite safe.

THERE were five peas in one shell. They were green, and the shell was green, and they thought all the world was green.

The shell grew and the peas grew, all sitting in a row. The sun shone and warmed the shell, and the rain made it clear and transparent.

"Are we to sit here forever?" asked one. "I'm afraid we shall become hard. It seems to me there must be something outside. I am sure of it."

Weeks went by; the peas became yellow, and the shell turned yellow. "All the world's turning yellow," said they; and they thought it was true.

Suddenly they felt a tug at the shell. The shell was torn off by some one's hands, and then put into the pocket of a jacket with other shells. "Now we shall soon be opened!" they said.

That is just what they were waiting for. "I should like to know who of us will travel the farthest!" said the smallest of the five.

" We shall soon know," said the eldest. " What is to be will be."

" Crack!" the pod burst, and the five peas rolled out into the bright sunshine. There they lay in a little boy's hand. He said they were fine for his peashooter, and he put one in and shot it out.

" Now I'm flying out into the wide world; catch me if you can!" and the first was gone.

" I shall fly straight into the sun," said the second.

" We'll go to sleep wherever we go," said the next two.

They were put into the peashooter, and, as they were shot out, said, " We shall go farthest."

" What is to happen will happen!" said the last.

As he was shot out, he flew against an old board under a garret window. Here was a crack filled with moss. The moss closed around him, making a soft bed.

In the garret there lived a poor woman who went out in the daytime to clean stoves and to do hard work.

She worked very hard, but she was still poor. Her sick daughter lived in the garret. This daughter was very weak, and for a whole year had kept her bed. This made her mother very sad.

The poor girl lay quiet all day long, while her mother went out to earn money.

It was spring. One morning as the mother was starting out, the sick girl looked through the lowest pane of the window.

"What is that green thing that looks in at the window? See, it is moving in the wind."

The mother stepped to the window and opened it. "Oh!" said she, "that is a little pea which has taken root here. See, it is putting out its leaves. How could it get into the crack? Here is a little garden you can watch."

The sick girl's bed was moved nearer to the window, so she could see the growing pea vine.

When the mother came home in the evening, the sick child said, "Mother, I think I shall get well. The sun shone in to-day very warm. The little pea vine is growing beautifully. I shall get better and go out into the warm sunshine."

The mother hoped this might be true, but she did not believe it would be so.

She put a stick into the ground, so that the wind might not break the vine. Then she tied a piece of string to the window-sill and the upper part of window. This was to give the pea vine something around which it could twine.

It seemed as if one could see it grow every day.

One day the mother said, "Here is a flower coming."

This made her very happy. She remembered that for some days her sick child had seemed brighter and happier. She had sat up in bed without her mother's help.

The child's eyes sparkled with delight when she saw the little flower.

A week later, the little girl sat up for a whole hour. She was very happy as she sat there in the warm sunshine. The window was opened, and just outside was a lovely pea blossom.

The sick girl bent down and gently kissed the pretty flower.

"The Heavenly Father has planted the pea and helped it to grow," said the happy mother. "It is a joy to you and to me."

But how about the other peas? The one that said, "Catch me if you can," as he flew out into the wide world, found a home in a pigeon's crop.

The two lazy ones were also eaten by pigeons.

The fourth, who wanted to go up into the sun, fell into the sink. It lay there for weeks and weeks, and swelled until it was very large.

" I am growing beautifully fat," said the pea. " I shall burst at last, and I don't think any pea could do more than that. I am the most wonderful of all the peas that were in the shell."

But the girl stood at the garret window, with bright eyes and the rosy hue of health on her cheek. She folded her thin hands over the pea blossom and thanked heaven for it.

Adapted from *Hans Andersen.*

11. A LAUGHING CHORUS.

O H, such a commotion under the ground
　　When March called, " Ho, there! ho!"
Such spreading of rootlets far and wide,
Such whispering to and fro.
And, " Are you ready?" the Snow-drop asked,
" 'T is time to start, you know."
" Almost, my dear," the Scilla replied;
" I 'll follow as soon as you go."
Then " Ha! ha! ha!" a chorus came
Of laughter soft and low
From the millions of flowers under the ground —
Yes — millions — beginning to grow.

" I 'll promise my blossoms," the Crocus said,
" When I hear the bluebirds sing."
And straight thereafter, Narcissus cried,
" My silver and gold I 'll bring."
" And ere they are dulled," another spoke,
" The Hyacinth bells shall ring,"
And the Violet only murmured, " I 'm here,"
And sweet grew the air of spring.

Then, " Ha! ha! ha!" a chorus came
Of laughter soft and low
From the millions of flowers under the ground —
Yes — millions — beginning to grow.

Oh, the pretty, brave things! through the coldest
    days,
Imprisoned in walls of brown,
They never lost heart though the blast shrieked
    loud,
And the sleet and the hail came down ;
But patiently each wrought her beautiful dress,
Or fashioned her beautiful crown ;
And now they are coming to brighten the world,
Still shadowed by winter's frown ;
And well may they cheerily laugh, " Ha! ha!"
In a chorus soft and low,
The million of flowers hid under the ground —
Yes — millions — beginning to grow.

Emerson's Evolution of Expression.

12. MOTHER NATURE'S BABY FERNS.

LAST autumn, when Jack Frost killed the mother fern, the poor little babies were much afraid. But some little snow-flakes came fluttering down from the sky and covered them with a warm blanket; and so they went to sleep.

They slept all through the long cold winter, and Jack Frost could not reach them.

Sometimes, in their dreams, they thought they heard old Northwind whistling over their heads. They did not care for that; the nice, white coverlet kept them so warm.

About the first of April, they were wakened by hearing voices all around them.

A sweet, old voice was saying, " Come, my darlings, it is time for you to get up."

Then such a scampering as there was; little flowers, grasses, and the baby ferns were all getting ready.

Mother Nature thought the ferns had better wrap up well, because some days the clouds might cover the sun, and then they would be very cold with old

Northwind for a playfellow. What do you think she gave them? Shall I tell you? She gave them little fur hoods.

They do not throw off these hoods the first warm day, but slip them back just a little at a time. They are afraid that old Jack Frost will catch them.

Old Mother Nature knows what a naughty fellow he is, so she told them just what to do.

When the sun has warmed the ground and air, they will help old Mother Nature make her woods beautiful.

13. MARSH MARIGOLD.

WHAT do you think the flowers did? I don't believe you could guess, if you were to try a week; so I shall have to tell you. They gave a party.

All the early flowers were invited, and they did such queer things! Miss Anemone had never seen Miss Marsh Marigold before, and she asked her to tell a short story about herself.

"Miss Anemone," said Miss Marsh Marigold, "my home is in a marsh. My good mother put me there that I might have plenty to drink, for I need a great deal of water.

"My home is always merry with music. Who do you think make all this music? No, it is not the birds alone. They sing in the morning, and Mr. Frog sings all the evening.

"When the warm sunshine called me, I sent up a strong, thick stem, with

smaller stems growing from it. The stems are not all alike. Some are called flower-stalks, and they are grooved. The others are the leaf-stalks. The leaf-stalks are not like the flower-stalks, for they have only one wide groove on one side.

"Little friend, if you are not tired I will tell you about my leaves. You see that I have some large leaves and some smaller ones. The smaller ones grow on the flower-stalks, and their stem is short.

"All my leaves are green, but they are darker on the upper side. They have no hairs on them as yours have, little Hepatica. They are very smooth."

As she said this, Miss Marigold turned to Miss Hepatica, who was standing close by. Then looking around at the other flowers, who had also drawn near to listen, she said: "I have something in my leaf which you all have. Can you guess what it is?"

"I know," cried little Hepatica.

"Then don't tell the rest," said Miss Marigold; "see if they can find out. I have a great many of them. They spread in my leaf, and carry food all over it."

"I know! I know!" cried several voices; "they are veins."

THE MARSH MARIGOLD.

"Yes," said Miss Marsh Marigold, "you are right. These veins start from the end of the leaf-stalk, and carry food to all parts. See how they divide again and again into veinlets!

"Do you not think my flowers are shaped like a saucer? Some of them are cup-shaped.

"There are five sepals in most of my flowers. The inside of the sepals is golden, but the outside of each is a pale green. Do you see how these sepals lap?

"I have many golden stamens, and some of my powder boxes have opened and let out the pollen. I saw little pollen fairies, dressed in gold, sailing away with the wind. They seemed so happy to be free.

"Then there are my pistils. They hold the seed boxes. Some of the flowers have five pistils; some six, and some even more.

"These little pistils are golden at the top, but below they are pale green. They are shorter and thicker than some of the pistils you have seen."

"Thank you," said Miss Anemone; "you were very kind to tell me so much about yourself."

"I am very glad if I have given you pleasure," said Miss Marsh Marigold.

14. MAY.

PRETTY little violets, waking from your sleep,
Fragrant little blossoms just about to peep,
Would you know the reason all the world is gay?
Listen to the Bobolink, telling you 't is May.

Little ferns and grasses, all so green and bright,
Purple clover nodding, daisy fresh and white,
Would you know the reason all the world is gay?
Listen to the Bobolink, telling you 't is May.

Darling little warblers, coming in the spring,
Would you know the reason that you love to sing?
Hear the merry children shouting at their play,
" Listen to the Bobolink, telling you 't is May!"

Elinor Smith's *Songs and Games for Little Ones.* OLIVER DITSON, Publ.

15. THE VIOLET.

A PURPLE violet, with her sisters, lived out in the woods under an oak tree. The violet had just unfolded her petals, and was very happy to be in the beautiful woods.

The oak tree kept the hot sun from her head and let in, only now and then, a sunbeam to warm her when the wind was cold. Day by day, the little violet grew larger and taller, and the bees came to get her honey.

They told her stories of the world outside, and she wished she could visit it. The violet could see

the children running to and fro. They laughed so merrily, she wondered what they were doing.

One morning the violet knew all about it, for a little girl, with some of her friends, ran up to her and cried, " Oh, what lovely purple violets for my May basket!"

Then away went the little blue violet, held tightly in the little girl's hand.

"Oh!" cried the little girl, "this pretty flower wears a purple bonnet. The five purple petals make this bonnet. It wears a collar of five green sepals."

" The lower petal has a spur or honey bag," said another. " I think the bees that visit the violet could tell you about the honey bag. I have heard that bees pay for all the honey they take. They do not know it, and it is done in a queer way."

" Yes," said the little girl, " the bee thrusts his head into the flower to reach the honey bag. He brushes against the pollen boxes, and some of the pollen sticks to him. Then, when he flies to the next flower, he rubs the pollen off on its knob. Do you know of what use this pollen is to the flower?"

" I have been told that seeds are much stronger
and better when they can get the pollen-dust from
another plant of the same kind," answered one of
her friends. " This is done by insects flying from
one flower to another and carrying the pollen on
their legs and bodies."

" Are the white and yellow violets your cousins
little violet?" asked one of the girls.

" Yes," answered the violet. " I am glad you are
going to take me with you, for now I can see the
beautiful world that the bees have told me about."

Before the violet had gone very far, the little girl
dropped her on the ground, and she lay there a
long time in the hot sun.

" If some kind person would only pick me up
and carry me home," she sighed.

Just as she said this, a little boy who was passing
said, " Oh, here is a poor violet that some one has
left in the hot sun to die." And he picked her up
and took her to his home.

Then the little violet was again happy in a pretty
vase in the little boy's room.

16. THE YELLOW VIOLET.

WHEN beechen buds begin to swell,
  And woods the bluebird's warble know,
The yellow violet's modest bell
  Peeps from the last year's leaves below.

Ere russet fields their green resume,
  Sweet flower, I love, in forest bare,
To meet thee, when thy faint perfume
  Alone is in the virgin air.

Of all her train, the hands of spring
  First plant thee in the watery mould,
And I have seen thee blossoming
  Beside the snow-bank's edges cold.

Thy parent sun, who bade thee view
  Pale skies, and chilling moisture sip,
Has bathed thee in his own bright hue,
  And streaked with jet thy glowing lip.

Yet slight thy form, and low thy seat,
  And earthward bent thy gentle eye,
Unapt the passing view to meet,
  When loftier flowers are flaunting nigh.

Oft, in the sunless April day,
   Thy early smile has stayed my walk,
But midst the gorgeous blooms of May,
   I passed thee on thy humble stalk.

So they, who climb to wealth, forget
   The friends in darker fortunes tried.
I copied them — but I regret
   That I should ape the ways of pride.

And when again the genial hour
   Awakes the painted tribes of light,
I 'll not o'erlook the modest flower
   That made the woods of April bright.

WILLIAM CULLEN BRYANT.

17. HEPATICA STORY.

IN the autumn, Mother Nature gave little Miss Hepatica to her nurse to be cared for and watched while Miss Hepatica slept all winter.

The nurse's name was Miss Underground Stem.

In the spring, Miss Hepatica was awakened by her playmate, Robin Red Breast, who called merrily to her with his " Cheer-up, Cheer, Cheer! "

Miss Hepatica was in a hurry, so Nurse Underground Stem sent servants in all directions through the ground to get the food and clothing which she would need in her upward journey.

It was still very cold above ground, and Miss Hepatica would need to be very strong and warmly wrapped, so that she would not take cold.

Nurse Underground Stem gave her a warm woolly dress and a hood which was tightly fastened, so that the cold March wind could not blow it off.

When Miss Hepatica had enough food to keep her alive and clothes to keep her warm in the

cold upper world, Nurse Underground Stem helped
her push the earth aside. At last her pretty little
head was above ground.

The sun warmed the ground, and the April rains
came and helped her grow. Robin Red Breast
sang cheerily from a tree near by.

Very soon she found her hood too warm, so she
unfastened it and let it fall back.

The sun and rain were very good friends to her.
She opened her little blue eyes to see her
old friends, Robin Red Breast and Old
Oak Tree. She nodded to all who passed,
she was so happy to see them after her
long sleep.

18. HEPATICA.

WHEN April awakens the blossom folk,
   And bluebirds are on the wing,
Hepatica muffled in downy cloak,
   Hastens to greet the spring.

Careless of cold when the northwind blows,
   Glad when the sun shines down,
She opens her wrap, and smiling, shows
   Her dainty lavender gown.

Her sisters are robed in pink, and some
   Are in royal purple dressed,
And over the hills and fields they come,
   To welcome the darling guest.

The children laugh as they pick the flowers,
   And the happy robins sing;
For, blooming in chill and leafless bowers,
   Hepatica means the spring.

Anna Pratt.

19. THE CARY TREE.

IN 1832, Alice Cary was twelve years old, and her sister Phoebe only eight. One day as these little girls were returning home from school, they found a small tree which a farmer had dug up and thrown into the road.

One of them picked it up and said to the other, "Let us plant it." These happy children ran to the opposite side of the road, and with sticks they dug up the earth.

In the hole thus made, they placed the little tree. They threw the earth around it with their little hands, and pressed it down with their tiny feet.

How they watched that tree to see if it were growing; and how they clapped their hands when they saw the buds start and the leaves begin to form!

They were happy all through the summer days watching it. When old Jack Frost and King Winter came, how they feared these rough fellows might kill it; and when the longed-for spring came, with what feelings of hope and fear did they go to find their tree!

When these two girls grew to be women, they moved to New York city, but they never returned to their old home without paying the tree a visit. They seemed to think as much of it as of their old friends.

That tree is now a large and beautiful sycamore. It is in the state of Ohio, and the people for miles around tell visitors how the sycamore was planted.

20. WHAT KATE HEARD THE FLOWERS SAY.

LITTLE Kate went out into the woods one sunny May afternoon to pick some flowers.

She had gone from flower to flower, gathering them until her basket was full. At last, she had walked a long way, and was very tired.

Setting her basket down beside her, she lay down to rest. In a few moments she heard soft music, and, listening closely, she found the sweet tones came from the basket beside her.

The violets were saying, "'We are as sweet as the roses and blue as the skies,' and we live in a soft mossy dell. The bees come and gather sweet honey from us."

" We," laughed the anemone, " live where we can frolic in the gentle breezes, and, because we love the wind, some people call us Wind Flowers."

The columbines said, " We live in that pleasant grove at the top of the hill. Our red and yellow dresses show very plainly above the green plants around us, and the children gather us in great bunches. The bees make us many visits, for we have five storehouses of sweet honey for them."

The bellworts now spoke in their bell-like tones: "We are wood flowers too, and our yellow bells hang ringing all day long. If it were not for our many leaves, which shut in the sound a little, I fear the other flowers would think us very noisy."

"Let us wake up the little girl now, for we are thirsty and need some water."

Kate sprang up as the bellwort rang. She had not been asleep, but had heard every word. How glad she was to have heard the flowers talk!

THE TRILLIUM.

21. THE TRILLIUM.

THIS queer flower is sometimes called the wake robin. It grows in shady places; sometimes you may find it in damp ones. Often its friend, the marsh marigold, is found but a short distance from it.

Its long green stem is round and smooth. Three large dark green leaves are at the top of the stem. The flower stem does not always grow straight up, but curves towards one side, so that the flower is often hidden by the leaves.

Most of the flowers you have studied have had five sepals. This flower has only three. Each is a light green, and has a pointed apex. These sepals stand out as straight as soldiers.

The three curved petals are snowy white, and, as they grow old, they wither.

This flower has fewer stamens than the earliest flowers of May, but they are very large.

The pollen box is also different in shape, being very long.

The pistil is the queerest of all the parts, being shaped like a vase. It is white, and has three little horns curving out at the top.

I wonder if any of the children have discovered why this flower is called a trillium? It is tri-leaved, tri-petaled, tri-everything.

22. THE DANDELION.

———◦———

"OH, Jessie!" said Emma, "look at this dear, little dandelion. It is the first one I have seen."

"Yes," said Jessie, "I think it is very beautiful, but in a short time there will be so many of its brothers and sisters that I fear you will not care for it."

"You are wrong, Jessie. If you look closely, I am sure you will think as I do, that it is a very beautiful flower. Its long slender leaves form a rosette on the ground. Mamma often has us

gather the leaves to cook.   They are called greens.
I like them, and am glad when we have them for
dinner."

"Did you ever think that when you have one
of these flowers in your hand, you have a whole
bouquet, — many flowers in one?"

"Yes," said Emma, "I have heard that people
call a flower that is made up of many flowers, a
compound flower."

"Let us look at one of these little flowers.   In
each I find petals, stamens, and pistil."

"The golden flowers of the dandelion are shut
up every night," said Emma.   "But in the morning
they are up much earlier than you or I, for they
open their eyes as soon as the sun is up."

"The dandelion has a curious habit," said Jessie.
"When the sun is very hot, it closes to keep from
wilting.   In this way, the green covering protects it
from the sun."

"Did you ever make curls of the hollow stems?"
asked Emma.   "The curls are pretty, but the milk
in them tastes bitter.   If we look at our hands after
gathering dandelions, we shall find that this bitter
milk has left dark stains."

"Oh, yes!" said Jessie, "and when the dandelions
have gone to seed, they look like white fluffy balls.

These seeds we have blown far away to see if we could tell the time of day."

"Old Mr. Wind carries these seeds far and near," said Emma. "Mother Nature cares for them during the long cold winter, and in the spring of the new year we shall find more of these little yellow flowers."

23. LEGEND OF THE DANDELION.

YEARS and years ago, a great many little stars
lived in the sky with their mother, the Moon,
and their father, the Sun.

Their mother liked to have them shine as soon
as it grew dark, to help brighten the sky, and so
make the earth lighter.

I do not know what could have happened to
these little stars, for they were usually very good
children; but one night when their mother called
them to come and light the sky, they came very
slowly, and when she told them to shine, they did
not do so.

They did just as I have seen some little children
do; they hung their heads and wore a cross look
on their faces.

Now, this made old Mother Moon feel very sad,
and when she saw that the children were not going
to do as she wished, what do you think happened?

Their places were taken by some good little
stars, who wished to help the people on the earth

find their way. Soon the naughty little stars felt
themselves falling, falling down from the sky.

Faster and faster they fell, till they sank down
into the earth.

The poor little things cried themselves to sleep,
because they were lonely, and because they were
so sorry that they had been naughty.

In the morning their father, the Sun, shone out
so brightly that everything wakened from sleep,
even the baby stars under the grass. When they
found themselves in the ground, they began to
cry again. The Sun heard them, and, seeing
how sorry they were, he called them up and
smiled upon them.

Now listen to what he told them. He said
that they were to shine on the earth instead of
in the heavens, and so make the earth beautiful.

So every morning, when the Sun smiled upon
them, they opened their eyes and shone all day.

At night, you can see the stars in the sky; and
every day in summer, the stars shining in the
grass.

Adapted.

24. THE DANDELION.

"BRIGHT little dandelion,
  Downy, yellow face,
Peeping up among the grass
  With such gentle grace;
Minding not the April wind
  Blowing rude and cold,
Brave little dandelion
  With a heart of gold.

Meek little dandelion
  Changing into curls
At the magic touch of these
  Merry boys and girls.
When they pinch thy dainty throat,
  Strip thy dress of green,
On thy soft and gentle face
  Not a cloud is seen.

Poor little dandelion,
  Now all gone to seed,
Scattered roughly by the wind
  Like a common weed.

Thou hast lived thy little life
   Smiling every day ;
Who could do a better thing
   In a better way ? "

From McMurry's *Classic Stories for the Little Ones.*
By permission of PUBLIC SCHOOL PUBL. CO.

25. THE ROBIN.

ON Washington's Birthday, Willie Clark had a holiday, and he thought he would have a fine time skating.

There were many boys on the pond. While racing with some of his friends, Willie fell and broke his leg.

Poor Willie! he had to stay in bed four weeks. That was a hard month for him; but, one bright morning the last of March, his father carried him to a window where there was a large easy chair.

"Thank you, father," he said. "This is very nice. Oh, look! the grass is green and the buds are opening. Hark! was that a bird singing?"

"Mother Nature has been hard at work while you have been in bed," said his mother. "There is a robin under the window. He is one of the first birds to come back to us, and we are always so glad to see him. We feel that when the robin comes, spring is coming too."

"Will you, please, bring me some crumbs, mother?" asked Willie, "I will throw them to him."

"Yes, my boy," said his mother. And she brought a plate of bread.

Willie fed the robin and found that the bird was not at all afraid.

"See, mother!" said Willie, "his tail, wings, and back are a grayish brown; his breast is chestnut-colored."

"Yes," said she. "Have you noticed his feet?"

"On each foot he has three toes in front and one behind. Is he a percher?" asked Willie.

"Yes," said his mother, "he belongs to the large family of perching birds."

"There he goes," said Willie. "See, how he runs over the ground! He does not hop as some birds do. What is he doing now? See him peck at the ground."

"He has found a worm and is pulling it out of the ground with his sharp bill."

As Mrs. Clark spoke, the robin flew into a cherry tree near the house.

"He must be building his nest in that tree," said Willie. "I know robins like to build their nests near houses. Let us watch and see if that is what he is doing."

Mr. and Mrs. Robin were very busy building a home. They made their nest of hay and grass fastened together with mud, and inside it was lined with fine soft grass.

When Willie was able to climb the tree, he found that Mrs. Robin had laid four pretty blue eggs.

26. THE NORTH STORY OF HOW THE ROBIN GOT
ITS RED BREAST.

LONG ago, in the far north, where it is very cold,
there was only one fire.

An old man and his little son took care of this
fire and kept it burning day and night.   They knew
that if the fire went out all the people would freeze
and the white bear would have the north land all to
himself.

One day the old man became very ill, so that his
son had everything to do.   For many days and
nights the boy bravely took care of his father and
kept the fire burning.   But at last he got so tired
and sleepy that he could no longer walk.

Now the white bear was always watching the
fire.

He longed for the time when he should have the
north land all to himself.

When he saw how tired and sleepy the little boy
was, he stayed close to the fire and laughed to himself

One night the poor little boy could keep awake no longer and fell fast asleep.

Then the white bear ran as fast as he could and jumped upon the fire with his wet feet and rolled upon it.

At last he thought it was all out, and went happily away to his cave.

But a gray robin had been flying near and had seen what the white bear was doing.

She waited until the bear had gone away.

Then she flew down and searched with her sharp little eyes until she found a tiny live spark.

For a long time she patiently fanned this spark with her wings.

Her little breast was scorched red, but she did not give up.

After a while a fine red blaze sprang up. Then she flew away to every hut in the north land.

Everywhere that she touched the ground, a fire began to burn.

So that soon instead of one little fire, the whole north land was lighted up.

Now, all that the white bear could do was to go further back into his cave and growl.

For now, indeed, he knew that the north land was not all for him.

And this is the reason why the people in the north country love the robin. And they are never tired of telling their children how it got its red breast.

Flora Cooke's *Myths.* A. FLANAGAN, Publ.

27. WHAT ROBIN TOLD.

HOW do robins build their nest?
　　Robin Redbreast told me.
First a wisp of yellow hay
In a pretty round they lay;
Then some shreds of downy floss
Feathers too, and bits of moss,
Woven with a sweet, sweet song,
This way, that way, and across:
That's what Robin told me.

Where do robins hide their nest?
Robin Redbreast told me.
Up among the leaves so deep,
Where the sunbeams rarely creep.
Long before the winds are cold,
Long before the leaves are gold,
Bright-eyed stars will peep and see
Baby-robins — one, two, three:
That's what Robin told me.

Geo. Cooper, in Little Flower Folks.

SPRING IN THE APPLE TREE.

28. SPRING IN THE APPLE TREE.

——•◦•——

ONE bright morning in spring, the sunbeams came down to visit the apple trees in the orchard. The leaves were putting on their green dresses, and the baby apple buds were just waking up. Some of the buds had on their dresses of pure white.

The buds had a happy time, for the bees and butterflies came to visit them, and the robins sang very cheerfully.

Mr. and Mrs. Robin had built a snug little nest among the branches, and they took very good care of the five little eggs tucked away under Mrs. Robin's warm wings.

One tiny cocoon cradle hung from a little twig near the robin's nest. Mrs. Robin said, "I hope my little birdies will wake up before that butterfly creeps out of its cocoon cradle; they will be so glad to see it.

" Mr. Robin and I worked very busily gathering tiny twigs and bits of hay and hair, and weaving them into a safe nest.

"We had a happy time building the nest, and we shall have a happy family in our little home when the birdies wake up."

Every day Mr. Wind came to sing among the apple-tree branches, and the sunbeams shone very warmly upon the little apple buds, and the cocoon cradle, and the birdies' nest. They called again and again, "Spring has come. Wake up! Wake up!"

One morning, Mrs. Robin heard a little cracking noise under her wing, and the eggs began to move. Then she heard a little voice say "peep" very softly.

What do you think had happened?

Yes, two little robins were waking up. They cracked the egg shells and put out two little heads. Four bright little eyes looked about with wonder at the beautiful white apple blossoms and the sunshine.

Very soon three more little robins awakened, and six more bright eyes looked and saw a very happy mother robin. While they all cuddled under her warm feathers, Mr. Robin flew about very busily, gathering food for his family of little birdies.

The little robins watched the cocoon cradle rocking in the breeze, and often said, "Mamma, what is that little thing swinging up there?"

Mother Robin said, "That is a baby's cradle; watch, and you will see the baby wake up and crawl out of the cradle. That baby learns to fly very quickly. When your wings have grown stronger, and your feathers have grown longer, I will teach you how to fly."

That made the little robins very glad, for they wanted to fly like Mother Robin.

As they watched the cocoon, one morning, they saw a little head peep out, and then — what do you think they saw next? A beautiful butterfly with golden brown wings.

They wanted to fly up to the butterfly, but they had not yet learned how to use their wings.

Mother Robin said, " To-morrow big Mother Tree wants every one to come to a party.

" Birds and bees are coming to sing; butterflies are coming; and the violets, crocuses, buttercups, and snow-drops that live down in the grass under the tree will be there with their smiling faces. You may fly down there to see them. O! we shall have a happy time."

The little birdies were so glad and talked so fast, they were not at all sleepy that night. They wanted the morning to come quickly.

The next morning, they were awake very early.

Mr. Wind carried all the invitations ; and when all the guests were gathered at the big apple tree, Mr. Wind played some soft music.

Each tried to make the others happy. The birdies sang and the flowers smiled. When the little robins succeeded in using their wings as Mother Robin showed them, they flew down among the flowers and spent a happy day.

They thought it a very beautiful world all about them.

Kindergarten News.

29. THE HUMMING BIRD.

A LICE and May were sisters. They had a large
yard in which to play, and loved to watch
John, the gardener, care for the flowers.

Alice thought it would be very nice if May and
she could have a flower bed of their own, and take
care of it without any help. May was much pleased
with this thought, and both children ran into the
house to find their mother.

" Mamma, may we have a garden of our own ? "
asked the girls.

"Certainly, my dears," answered their mother, "if
you will give the plants good care."

The girls promised, and ran into the yard to
select a spot with plenty of sunshine.

The gardener gave them a great
many plants, and in a few weeks they
had beautiful flowers of all colors.

One day, when they were caring
for their flower friends, they saw what seemed
to be a flash of sunshine. They watched it,

and found that instead of sunshine, it was a little bird.

"Oh, May," said Alice, "what a beautiful bird it is! See, it stays right over that flower! It does not rest on the flower. It is held up by its fluttering wings. Its wings never seem to be still."

"Yes," said May, "let's call John and see if he can tell us its name."

"John, please come here and see if you can tell us the name of this bird," called Alice.

John was working near, and came hurrying to them. When he reached the girls, the bird had gone, but in a minute, it came back to the flowers.

"That is a humming bird," said John. "It comes here every day to make your flowers a visit. The Indians call it 'a living sunbeam.'"

"But why does it come here?" asked May.

"Don't you see that it runs its long bill down into the flowers and gets the sweet honey they hold? In the same way, it takes any little insects that may be in them."

"There it goes!" said Alice. "Hear it hum!"

"It makes that noise with its wings," said John. "It flies so swiftly that we can hardly see that it has wings."

HUMMING BIRDS.

" Have you ever seen a humming bird's nest ? " asked Alice.

" Yes," said John, "last summer two humming birds had a nest in that tree."

" Please tell us all about it," said Alice.

" Well, the nest was a little fairy cradle, about half the size of a hen's egg and as soft as cotton. It was made of soft moss and lined with the down of the young fern and soft cotton fibers. The father bird brought the materials for the nĕst, and the mother bird arranged them.

" When I found the nest, it had two little eggs in it. Each was a trifle larger than a pea and quite white.

" A little later, I saw two tiny birds, no larger than bees. You must be very careful when you go near a nest, for the mother bird fears you will take her babies. She will fly at you and try to pick out your eyes.

" Not long afterwards, I saw the father and mother birds teaching the little humming birds to fly, and soon they were gone."

The girls were much pleased with John's story. They came every day, while the flowers lasted, to watch their new friend.

30. THE WOODPECKER.

ONE bright morning in April, Alice and Mabel were out walking near a wood. Suddenly they heard a tap, tap, tap-tap-tap!

"What is knocking?" asked Alice.

"It comes from this large old oak," answered Mabel.

"Oh, it is a bird! I see his red head. Let us watch him."

"His breast is a yellowish white, but his back is speckled, black and white," said Mabel.

"See, there he goes! How fast he climbs up the tree! I should think he would fall off," exclaimed Alice. "His feet cannot be like Mrs. Robin's."

"No," said Mabel. "This bird is a climber, and they have two toes in front and two behind. Their claws are long and curved."

"I wonder how he makes that noise," said Alice.

" He makes it with his bill. It must be very sharp, for he sticks it way into the wood," said Mabel.

" Why is he tapping ? " asked Alice.

Their mother, who was not far behind, came up just in time to answer this question.

" He is eating his breakfast, Alice."

" Out of wood ? How can that be ? " asked both girls at once.

Mamma stepped to the old tree, and stripped off a piece of bark. " What do we find here, girls ? "

" Little worms and bugs," said Alice.

" Those are what Mr. Woodpecker is after. There is something very strange about his tongue. The end of it is covered with little sharp points," said mamma, " so the insects cannot get away from him."

" Mamma," said Alice, " some of the insects are very tiny."

" Yes, but he has another way of catching those. He has a sticky substance on the end of his tongue, and the tiny insects are caught by it."

" Look just above your head, dear. Do you see the little round hole in that large dead limb ? "

"Yes, but where does Mr. Woodpecker build his nest?" said Mabel.

"Ha! Ha!" laughed mamma, "that is the door to his little house in the limb. He and his mate take turns in digging with their sharp bills. They make the hole larger and larger, until they have a little hall and a room down stairs. They often make a deep hole. It is sometimes fifteen or twenty inches deep.

"Don't you think the little birdies are well sheltered from the rain and wind in that deep room?" asked their mother.

"Yes, indeed," answered the girls, "he takes good care of his little ones."

THE WOODPECKER.

31. THE ORIGIN OF THE WOODPECKER.

THERE was an old lady who lived on a hill. She was very small, and she always wore a black dress, and a large white apron with big bows behind.

On her head, she wore the queerest little red bonnet you ever saw.

It is a sad thing to tell, but the little old lady had grown very selfish as the years went by.

People said this was because she lived all alone, and thought of no one but herself.

One morning, as she was baking cakes, a tired, hungry old man came to her door.

"My good woman," said he, "will you give me one of your cakes? I am very hungry. I have no money to pay for it, but whatever you first wish for, you shall have."

Then the old lady looked at her cakes, and thought that these were too large to give away.

So she broke off a small bit of dough, and put it into the oven to bake.

But when it was done, she thought this one was too nice and brown for a beggar.

So she baked a smaller one, and then a smaller one, but still each was as nice and as brown as the first.

At last, she took a piece of dough only as big as the head of a pin, yet even this, when it was baked, looked as large and fine as the others.

So the old lady put all the cakes on the shelf, and offered the old man a dry crust of bread.

But the poor man only looked at her, and before the old lady could wink her eye, he was gone.

Then the old lady thought a good deal about it, and knew that she had done wrong.

" Oh, I wish I were a bird," said she, " I would fly to him with the largest cake on the shelf."

As she spoke, she felt herself growing smaller and smaller, until the wind picked her up and carried her up the chimney.

When she came out, she still had on her red bonnet and black dress. You could still see her large white apron with the big bows behind.

But she was no longer an old lady, but a bird, just as she had wished to be.

But she was a wise bird and cheerfully began to pick her food out of the hard wood of a tree.

And people after a while, when they saw her at work, named her the red-headed woodpecker.

Flora Cooke's *Myths.* A. FLANAGAN, Publ.

32. THE DUCK.

THE duck is a swimming bird. It can swim, dive, and walk. It does not walk well, and instead of saying a duck 'walks,' we often say it 'waddles.'

Its short, strong legs are set far back so that it can push itself through the water.

·The duck's feet have three toes in front, which are united by a web. The hind toe is short. When this bird swims, it uses its webbed feet as paddles.

It has a boat-shaped body, which is covered with feathers. Underneath the feathers is a thick covering of down.

Does the duck get wet when it swims?

Oh, no! the down and the oiled feathers keep it dry. Did you ever see one make its toilet?

It seems to know that its feathers overlap one another so closely that when they are all straight and well oiled, the water cannot reach its body.

All birds keep their feathers in very neat order. I do not remember ever to have seen one dirty or untidy, unless it was too ill to care for itself.

The duck's bill is broad and flat. The duck has no teeth; but there are little notches on the edges of the bill. This bird is very fond of gathering some of its food under the water. These notches are used to drain the water from the food it gets in this way.

Have you ever thought that the flying and swimming birds use their tails as rudders?

The flying birds steer themselves through the air by their tails; the swimming birds through the water.

Do you know the names of other birds belonging to the family of swimmers?

33. THE HEN.

ONE day in summer, all the fowls in Farmer Green's barnyard had a meeting. Each, in turn, was to tell something about himself. It came the hen's turn to talk.

"My friends," she said, "I belong to the family of scratchers. My legs are short and are in the center of my body. On each foot, I have three toes in front and one behind.

"All my family have rough toes and nails that are short and blunt. If our nails were long and sharp, they would not help us in scratching, and we could not walk so well. When we place our feet on the ground, the toes spread.

"Our home is not in the air or in the water, but on the ground. So our wings are not strong.

"My sister hens and I are never idle, but scratch all day for our food. We like very much to eat the worms and bugs we find. I have heard children say, 'I wonder how the hen can see those little bugs, I cannot.' They

FARMER GREEN'S BARNYARD.

forget our legs are so short that our sharp eyes are only a little way from the ground.

"Sometimes, we take time to go into the barn where Farmer Green has made our nests, and there we lay nice fresh eggs.

"My sister, Mrs. Speckle, has been sitting on some eggs for about three weeks. To-day, she came off the nest with her little family.

"The little chicks look like fluffy balls on legs.

"Mrs. Speckle scratches until she finds a nice bug or worm, and then calls, 'Cluck, cluck,' which means 'Come here, children, I have found something for you.' How quickly they snatch up the worm or bug!

"We do not have to feed our babies as Mrs. Robin does. At first, we scratch for them, but very soon, they do this for themselves. It is very funny to see young chicks try to scratch with their tiny feet.

"Mrs. Duck asks if we have teeth to grind our food. No, we have no real teeth, but we have something which takes their place.

"Our food passes from the mouth into the crop, and from the crop to the gizzard. Here, the grinding takes place.

"When we are picking up bugs and worms, we also pick up gravel which helps to grind our food.

"My friends, I think I will let some one else take my place. I have talked for a long time."

THE SNIPE.

34. THE SNIPE.

THE snipe makes its home in damp, marshy places. It belongs to the family of wading birds.

Many of the members of this family have long legs, but the snipe's legs are short. With its long, straight, slender bill, it feels about in the mud for worms.

Its eyes are large, and placed far back to guard it against danger when it is feeling for food in the mud.

Each foot has three toes in front, and one behind. The tail is short, with either fourteen or sixteen feathers.

The nest is made of dry grasses, in a small hole in the ground, or in a tuft of grass or rushes. The eggs are greenish-yellow speckled with brown.

The flesh of the snipe is highly prized for food, but the young hunter finds it a hard bird to shoot, for when frightened, it will fly in a zigzag course through the air.

The snipe is very fond of its young. If it hears you coming near its nest, it will keep perfectly still, hoping that it will not be noticed.

It will not move until you have reached its nest. Then it will fly off a short distance, and cry, and limp as if it were hurt. It hopes you will feel sorry for it, and come to help it, leaving its young undisturbed.

35. JACK AND THE OSTRICH.

ONE afternoon, Jack was all alone in the sitting room. It was very quiet, and he was looking at a book containing pictures of birds. He came to a picture of an ostrich, and thought "What a queer-looking bird that is!"

Soon his eyes began to close, his head nodded, and Jack was fast asleep.

Do you know where 'the Land of Nod' is? Well, Jack found himself in 'the Land of Nod.' What do you think he saw? A large ostrich just like the one in his book.

"How do you do, little boy?" said the ostrich.

"I am well, thank you. How are you?" said Jack. "Where did you come from, Mr. Ostrich?"

"My home is across the sea in the hot desert," said the ostrich.

"How tall you are!" said Jack.

"Yes," said the ostrich, "I am the largest of all the birds. I am seven feet high."

"Oh! your head is higher than my father's. Why do you need that long neck?"

"Don't you know, little boy, I have to reach up into the tall trees to get some of my food?"

"I thought you ate glass, stones, and leather," said Jack.

"I do, but only when I can find nothing better. I often swallow small stones to help grind my food, but I am fond of fruit, leaves, grain, snails, and insects."

"You have a very small head and wings. Can you fly, Mr. Ostrich?" asked Jack.

"Oh, no! my wings are too small; but I can run faster than any horse. Would you like to see me?"

Before Jack could answer the ostrich was gone. How fast he ran! In a few minutes Jack could not see him.

"Oh!" said he, "I do hope he will come back, for I like to talk with him."

THE OSTRICH.

Jack waited patiently, and soon saw the ostrich running back towards him.

"Well, Mr. Ostrich," said Jack, "that was fine. I should not care to race with you. Can you tell me how you run so swiftly?"

"Certainly," said the ostrich, "look at my feet, and you will see a pad under each. These help me to spring at every step. My wings help me, and my legs are so long that I get over the ground very rapidly.

"Jack, did you ever see any ostrich's eggs?"

"No," said Jack, "please tell me about them."

"Well, Mrs. Ostrich and two or three of her friends dig a hole in the sand. The mothers lay their eggs in the same hole, and take turns in sitting upon them. Sometimes, one of the fathers takes a turn sitting, while the mothers go visiting. Then the hot sun shines on the eggs every day and helps them to hatch.

"The eggs are as large as cocoanuts. People, who make their homes in the desert, eat these eggs. One egg is large enough for the whole family."

"Your head and neck are nearly bare, and your prettiest feathers grow in your wings and tail," said Jack.

" Yes," said the ostrich, "our wing and tail feathers are the plumes that the ladies wear in their hats. I must go now. Good-bye, Jack."

Just then Jack heard a noise. He awoke with a start. " Well," said he, " I have had a fine talk with Mr. Ostrich, and learned many things about him.

" My book tells me a queer thing about these birds. When they wish to hide, they bury their heads in the sand. They seem to think that they cannot be seen, because they cannot see. They are often caught because of this."

36. THE OWL.

THE owl is one of those birds which go about
in the nighttime in search of food.   It is
called a bird of prey, because it lives on birds not
so strong as itself, mice, and other small animals.

Owls are often called winged cats, or cats in
feathers.   Tame owls become very friendly with
puss, just as if they knew that she was a relation.

During the daytime, owls hide away in holes
in trees, in caves, and old buildings, but in the
dusk of the evening, when they can see better than

in the broad daylight, they fly around looking for game.

They catch their prey with their claws, and if it is small, they swallow it whole; the bones, hair, or feathers being afterward thrown up rolled in a ball. If it is large, they first tear it into pieces.

Owls' feathers are very soft, so their flight is noiseless. The feathers, even those of the wings, are downy.

The legs and feet, of most owls, are feathered to the toes, and in many kinds, even to the claws.

Owls have large heads with flat faces. The eyes are round and staring. They look very wise and knowing.

The bill is sharp and hooked, curving almost from its base.

During the night, they hoot sometimes for hours. Often the noise sounds so much like a person in distress that people, who do not know them, have gone out with lanterns to see what was the matter. All the time, the owls have been enjoying themselves, hunting mice and birds.

The barn owl lives in old buildings, or the hollow trunks of trees. It destroys great numbers of rats and mice, and is a true friend to the farmer.

THE OWL.

If the barn owl catches more food than it needs, it stores it up for future use.

It hardly ever leaves its home by day, unless driven out. When this is the case, all the little birds in the neighborhood attack it in flocks. They seem to think it an enemy which they know they can annoy. As the owl cannot see very well, it sometimes has a hard time defending itself.

The barn owl has perfect disks of feathers around the eyes. The tail is rather short and round; the wings reach beyond the tail. The toes of this owl are not feathered.

The horned owls have tufts of feathers on their heads which stand up like horns or ears.

If these tufts are pushed away, you can see a curious opening into the head which is the true ear.

The screech owl is much like the horned owl, only smaller.

37. THE PIGEON AND THE OWL.

THERE once was a pigeon, as I have heard
    say,
    Who wished to be wise;
She thought to herself, "I will go to the owl,
    Perhaps he'll advise;
And if all he tells me, I carefully do,
I'll surely get wisdom." Away then she flew.

When little Miss Pigeon arrived at the barn
    She found the owl there;
Most humbly she cooed out her wish, but the owl
    Did nothing but stare.
"Well, well," thought Miss Pigeon, "of course I
    can wait,
I won't interrupt him; his wisdom is great."

She waited and waited. At last the owl blinked,
    And deigned a remark,
"You'll never be wise, foolish pigeon, unless
    You stay in the dark,

And stretch your small eyes, and fly out in the
    night,
And cry, ' Hoo-hoo-hoo,' with all your might."

So little Miss Pigeon to practice began;
    But all she could do
Her eyes would not stretch and her voice would
    not change
    Its soft, gentle coo;
And she caught a sad cold from the night damp
    and chill,
And lacking the sunshine besides, she fell ill.

Then little Miss Pigeon gave up being wise;
    " For plainly," said she,
" Though owls are the wisest of birds, theirs is not
    The wisdom for me;
So I 'll be the very best pigeon I can."
And what do you think? She grew wise on
    that plan.

From McMurry's *Classic Stories for the Little Ones.*
By permission of PUBLIC SCHOOL PUBL. CO.

38. THE FROGS' EGGS.

ALBERT had gone to visit his uncle who lived in the country. One day in April, when they were taking a walk near a pond, Albert saw on the surface of the water something which was strange to him.

"Uncle George," said he, "what is that jelly-like mass floating on the water? It has black dots in it."

"Those are frogs' eggs," said his uncle. "They are sometimes called frogs' spawn. Let us get some of them in a pail, and take them home."

They took the eggs home, and put them into a tub of water. In two or three days, they noticed that the little eggs became larger and larger; and instead of remaining round, they became oblong.

About a week later, when Albert looked into the tub, he cried, "Oh uncle, there are some little fish swimming around. They look all head and tail!"

"Some of the eggs have hatched, and those are little tadpoles. Now, you must go to the pond and get some pond-weeds for them to eat."

" Yes, I will go; but will the tadpoles ever be frogs?" asked Albert.

"Watch them, and see what you think about it," said his uncle.

Albert did as his uncle told him, and this is what he saw.

Soon, some gills, which the tadpoles used in breathing, grew outside the head.

When the tadpoles had grown a little more, the hind legs began to develop. The gills grew smaller. Then the front legs appeared, and the tails became shorter. The gills were covered by a thin skin, and changed into lungs.

Then, much to Albert's delight, three little fellows hopped out of the tub, — little frogs with tails. But these tails grew smaller and smaller, and at last disappeared.

Albert knew that the frogs would be happier near a pond, so he took them to one near by, where many of their relations were croaking.

39. FROGS AND TOADS.

ALBERT stayed to watch the frogs. Soon, he saw their cousin, the toad, come hopping along. What a merry family they were!

The bullfrog would give a deep, gruff "Ker-chog! ker-chog!" and the smaller frogs would answer in pleasanter tones.

Just then, his uncle came to the pond and said, "Let me tell you something about the frogs and toads:

"Frogs spend most of their time on land now. Frogs and toads breathe by taking air into their mouths, and swallowing it. That is what they are doing when you see their throats swelling out.

"They have another way of breathing. There are little pores in their skin through which they breathe. This is one reason why they like moist, shady places when the sun is shining. The hot sun dries up the pores.

"The frog's skin is smooth, but the toad is covered with warts. People used to believe

THE FROGS.

that handling toads would give one warts, but that is not true.

"The frog's foot has four toes or fingers in front, and five toes behind. The hind legs are very long, and are used for leaping and swimming.

"The hind toes are united by a web which helps the frog in swimming.

"The tongue of the frog or toad is very curious. It is fastened at the front of the jaw. It is long, and he can roll it out very quickly. You cannot see him do it. The tongue is covered with a sticky substance, which helps to hold the insects that he catches.

"Toads are useful in gardens, for they eat many insects which would do great harm by destroying the flowers and vegetables.

"Gardeners feel very friendly toward the toad, and are always glad to see him in their gardens. Although toads eat so many insects, they can go a long time without eating. Toads can easily be tamed, and people often make pets of them.

"The mother toads go to the ponds to lay their eggs, and the toads' eggs are hatched in the same way as the frogs' eggs.

"During the winter, some frogs and toads bury themselves in the mud in the bottom of ponds."

40. THE TURTLES.

CHARLES had a cousin named Roy, who lived in the country, near a small lake. Many turtles made their home in this lake.

Roy came to visit his cousin at the latter's city home. He brought him two baby turtles in a tin can.

Charles was much pleased with his new pets, and the boys set to work to make the turtles a home. They found an old tub, and put in water and stones. Sometimes, the baby turtles would swim about; at other times, they would crawl up on the stones.

The boys caught flies for the turtles to eat. How they laughed to see the little turtles snap them up!

One day, as they sat watching their pets, Charlie said, "Roy,

you have lived near turtles so long, tell me something about them."

"There are some kinds of turtles I have never seen; but I have heard and read about them," said Roy.

"Some live on the land, and are called land turtles; some live in the sea, and are called sea turtles; some live in lakes, rivers, and streams, and are called river or pond turtles.

"The turtle has a hard shell for its house. It draws in its head, legs, and tail when danger is near."

Charles had noticed this, and took one of the turtles in his hand to count the plates on the top of its house. How many large ones did he find? How many small ones around the edge?

"Put the turtles on the floor," said Roy, "and watch them walk. See how they push themselves along with their sharp claws! Hear their shell go rap, rap, as they move along. Can you tell why it makes this rapping noise?

"I once had a turtle, which I kept in the house all winter. It became very tame, and would follow me all over the lower part of the house.

"It could not go up stairs, but one day it

followed me down. I had taken it up stairs and left it while I went down to read.

"Soon, I heard a noise. It sounded as if something had dropped on the stairs.

"Going out into the hall, I saw the turtle on one of the top steps. I stood still to see what it would do.

"It crawled out until its shell just balanced on the edge of the step. Then it drew in its head, legs, and tail, and dropped to the next step. It waited a minute, then out came its head, legs, and tail, and it crawled to the edge of that step. Here, it balanced itself ready for its fall to the next. It kept this up until it came down all the stairs to the place where I stood.

"Turtles have been known to go without anything to eat for weeks, and it has not seemed to harm them. They have no teeth; but their jaws are covered with horn.

"Most land and pond turtles are quite harmless, if you are careful not to put your fingers near their mouths. But there is one kind of freshwater turtle that would bite off your finger in a moment. This is called the snapping turtle. The snapping turtle lives in the water most of the time, but comes out long enough to lay its eggs.

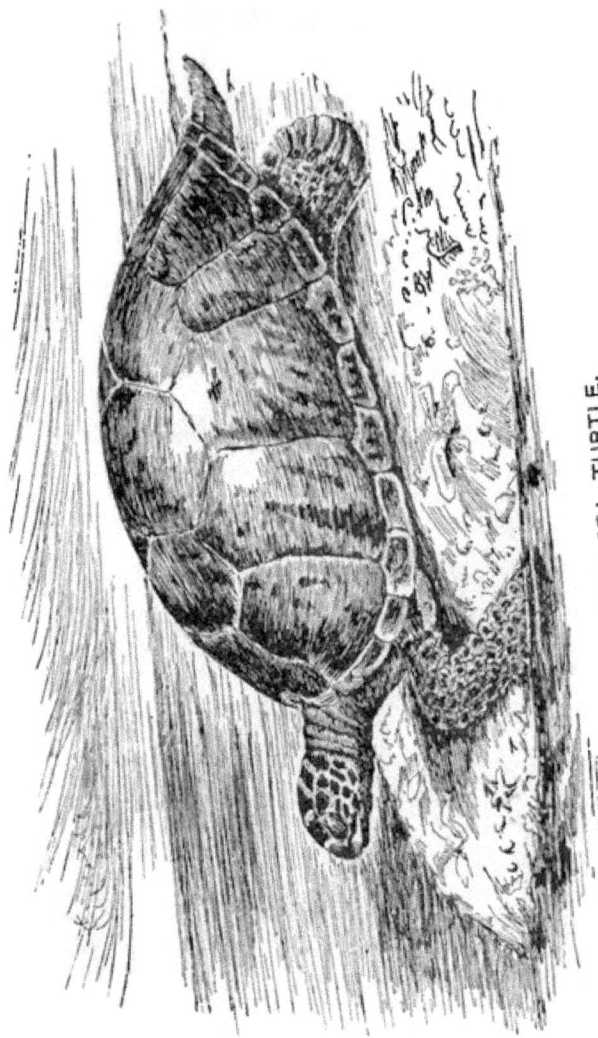

THE SEA TURTLE.

"This turtle grows to be very large. Some have been found more than four feet long. Its shell is so hard, that it will bear the weight of a man. The flesh of snapping turtles is good for food, and they are caught to sell.

"The sea turtle comes out of the water only long enough to lay its eggs. Its legs are made for swimming, and look like paddles. Its fore legs are much the longer. The legs are so very strong, that the turtles can move through the water like a flash.

"In the warm days of spring, the turtle lays its eggs in the dry sand. It digs a little hole with its feet, and drops in the eggs. It scratches and smooths the sand over them, so that few people would notice that they had been laid there.

"It does not stay to hatch its eggs, but leaves that for the warm sun to do. When the eggs are hatched, the baby turtles scratch their way out of the sand and crawl off.

"The turtles do not like King Winter with his snow and ice; so they bury themselves in the sand or mud. There they sleep until he has gone back to his home in the far north."

ADVERTISEMENTS

CYR'S CHILDREN'S READERS.

By ELLEN M. CYR.

A series of readers prepared expressly for the first, second and third years of school life. These books are pervaded with the spirit of child-life, and all the best devices and methods are made use of to render the first steps in reading easy, interesting, and judicious.

THE CHILDREN'S PRIMER. Fully illustrated. Sq. 12mo. Cloth. 96 pages. Introduction price, 24 cents.

THE CHILDREN'S FIRST READER. Fully illustrated. Sq. 12mo. Cloth. 101 pages. Introduction price, 28 cents.

THE CHILDREN'S SECOND READER. Fully illustrated. Sq. 12mo. Cloth. 186 pages. Introduction price, 32 cents.

THE CHILDREN'S THIRD READER.   [*See Announcements.*

READING SLIPS. To be used in connection with The Children's Primer, or independently, or with any other first reading book. Forty-eight manilla envelopes containing each twenty sentences, printed in large type on stiff manilla paper. Introduction price, five cents per envelope.

The **Children's Primer** is written especially for the children, and, as one may say, *with* the children. It contains more reading-matter, in proportion to the number of new words, than any other book in the market.

The sentences are short, childlike, and full of expression ; the illustrations, not only artistic, but suggestive ; and the new words are introduced so gradually and repeated so constantly that the little ones are enabled to hold fast all they learn.

The **Children's First Reader** is made for the second half-year. It is a simple but steady growth in the same line with the Primer, and makes possible a real advance instead of a constant stopping and beginning over again.

The **Children's Second Reader** is written on the same general plan of slow but steady progress manifested in the first books of this series.

Stories from the lives of Longfellow and Whittier form a considerable share of its contents. These bear especially upon the relation of the two poets to child-life, and are intended to awaken a personal interest in them and their poems.

THE JANE ANDREWS BOOKS.

By Jane Andrews.

A remarkable series of attractive and interesting books for young people, — written in a clear, easy, and picturesque style. This is the famous Jane Andrews series, which has been for many years an old-time favorite with young folks.

THE SEVEN LITTLE SISTERS WHO LIVE ON THE ROUND BALL THAT FLOATS IN THE AIR. Cloth. 143 pages. Illustrated. For introduction, 50 cents.

EACH AND ALL; THE SEVEN LITTLE SISTERS PROVE THEIR SISTERHOOD. Cloth. Illustrated. 162 pages. For introduction, 50 cents.

THE STORIES MOTHER NATURE TOLD HER CHILDREN. Cloth. Illustrated. 161 pages. For introduction, 50 cents.

TEN BOYS WHO LIVED ON THE ROAD FROM LONG AGO TO NOW. Cloth. 243 pages. Illustrated. For introduction, 50 cents.

GEOGRAPHICAL PLAYS. Cloth. 140 pages. For introduction, $1.00.

Books of genius are rare, and rarest of all among books for schools. **The Jane Andrews Books** appear to be of this remarkable sort. They are not to be classed with the good but commonplace things that any of us might write on the same subject. There is found about them the certain indefinable *something* which, as the ancients said of the walk of their divinities, reveals more than ordinary mortal. As a consequence they possess a peculiar quality of life and interest and inspiration. They not only instruct, but quicken.

The "Seven Little Sisters" represent the seven races. The book shows how people live in the various parts of the world, what their manners and customs are, what the products of each section are and how they are interchanged.

"Each and All" continues the story of "Seven Little Sisters," and tells more of the peculiarities of the various races, especially in relation to childhood.

In "Stories Mother Nature Told" Dame Nature unfolds some of her most precious secrets. She tells about the amber, about the dragon-fly and its wonderful history, about water-lilies, how the Indian corn grows, what queer pranks the Frost Giants indulge in, about coral, and starfish, and coal mines, and many other things that children delight to hear.

In the "Ten Boys" the History of the World is summarized in the stories of Kabla the Aryan boy, Darius the Persian boy, Cleon the Greek boy, Horatius the Roman boy, Wulf the Saxon boy, Gilbert the Knight's page, Roger the English boy, Fuller the Puritan boy, Dawson the Yankee boy, and Frank Wilson the boy of 1885.

The able, suggestive, and interesting "Geographical Plays" is designed as a sort of review of each country or topic, and it presents a comprehensive view of the subject as a unit. It is used after a country has been faithfully studied from the geography, and when the pupil has become familiar with all names given in the play. These plays are well written, and are calculated to produce an animating effect upon a school.

E. A. Sheldon, *Principal State Normal School, Oswego, N.Y.:* They are excellent. I know of nothing equal to them for the purpose for which they have been prepared. We have long used them in our own school of practice, and can recommend them most heartily.

John G. Whittier: I have been reading the new book by Jane Andrews, "Ten Boys." In all my acquaintance with juvenile literature I know of nothing in many respects equal to this remarkable book.

Mary E. Burt, in "*Literary Landmarks*": I have seen a six-year-old girl read Jane Andrews' "Seven Little Sisters," and "Each and All," repeatedly, with renewed interest at each reading.

Thomas W. Higginson: I think that the mere reading of this book "Seven Little Sisters" — read over and over, as children always read a book they like — will give to the young readers a more vivid impression of the shape of the earth, and of the distribution of nations over it, than the study of most text-books.

OPEN SESAME.

ABOUT ONE THOUSAND PIECES OF THE CHOICEST PROSE AND VERSE. Compiled by BLANCHE W. BELLAMY and MAUD W. GOODWIN. In three volumes. Sq. 12mo. About 350 pages each. Illustrated. For introduction, 75 cents each.

Among the many collections of choice extracts, this is distinguished for its comprehensiveness, the care and skill with which the work has been done, the gradation of the pieces, their topical arrangement, and the mechanical excellence of the volumes, — clear, large type, fine illustrations, and handsome binding.

PHYSIOLOGY AND BOTANIES.

A HYGIENIC PHYSIOLOGY. For the use of schools. By D. F. Lincoln, M.D. Cloth. Illustrated. 206 pages. For introduction, 80 cents.

The chief object of this book is to present the laws of health as fully as pupils fourteen or fifteen years old can understand, appreciate, and apply them.

LITTLE FLOWER-PEOPLE. By Gertrude Elisabeth Hale. Sq. 12mo. Illustrated. Cloth. 85 pages. For introduction, 40 cents.

The aim of this book is to tell some of the most important elementary facts of plant life in such a way as to appeal to the child's imagination and curiosity.

GLIMPSES AT THE PLANT WORLD. By Fanny D. Bergen. Fully illustrated. Cloth. 156 pages. For introduction, 50 cents.

This is a capital child's book, for it covers quite an area of botanical study and presents a good array of interesting facts relative to plant life.

OUTLINES OF LESSONS IN BOTANY. For the use of teachers, or mothers studying with their children. By Jane H. Newell. **Part I.: From Seed to Leaf.** Square 12mo. Cloth. 150 pages. Illustrated. For introduction, 50 cents. **Part II.: Flower and Fruit.** 393 pages. Illustrated. Cloth. For introduction, 80 cents.

A READER IN BOTANY. Selected and adapted from well-known authors. By Jane H. Newell. **Part I.: From Seed to Leaf.** Cloth. 199 pages. For introduction, 60 cents.

The purpose of this book is to supply a course of reading calculated to awaken the interest of the pupil in the study of the life and habits of plants.

A READER IN BOTANY. Selected and adapted from well-known authors. By Jane H. Newell. **Part II.: Flower and Fruit.** Cloth. 179 pages. For introduction, 60 cents.

This volume follows the first part in supplying a course of reading in Botany for pupils of the higher grades. It deals with the life-habits of plants, especially as relating to the flower and fruit.

www.ingramcontent.com/pod-product-compliance
Lightning Source LLC
Chambersburg PA
CBHW021933190326
41519CB00009B/1007